THE POETRY OF MERCURY

The Poetry of Mercury

Walter the Educator

Silent King Books

SILENT KING BOOKS

SKB

Copyright © 2024 by Walter the Educator

All rights reserved. No part of this book may be reproduced in any manner whatsoever without written permission except in the case of brief quotations embodied in critical articles and reviews.

First Printing, 2024

Disclaimer
This book is a literary work; poems are not about specific persons, locations, situations, and/or circumstances unless mentioned in a historical context. This book is for entertainment and informational purposes only. The author and publisher offer this information without warranties expressed or implied. No matter the grounds, neither the author nor the publisher will be accountable for any losses, injuries, or other damages caused by the reader's use of this book. The use of this book acknowledges an understanding and acceptance of this disclaimer.

"Earning a degree in chemistry changed my life!"
— Walter the Educator

dedicated to all the chemistry lovers, like myself, across the world

MERCURY

Where shadows dance, a liquid gleam,

MERCURY

Mercury, elusive, in the alchemist's dream.

MERCURY

A metal of mystique, a liquid silver stream,

MERCURY

In the cosmos' tapestry, a shimmering seam.

MERCURY

From ancient lore to modern science's glance,

MERCURY

Mercury's tale weaves, in a celestial trance.

MERCURY

A messenger of gods, in Roman expanse,

MERCURY

Swift-winged sandals, in Hermes' advance.

MERCURY

In thermometers, it rises and falls,

MERCURY

A silvery measure within glass walls.

MERCURY

A conductor of heat, as nature calls,

MERCURY

In alchemical rituals, it enthralls.

MERCURY

But tread with caution, for Mercury's embrace,

MERCURY

Can lead to peril in its toxic space.

MERCURY

A clandestine poison, without a trace,

MERCURY

In rivers and soils, leaving its base.

MERCURY

Yet in its essence, a paradox found,

MERCURY

In art and science, its wonders abound.

MERCURY

A mirror to stars, with allure unbound,

MERCURY

In planetary cycles, its dance profound.

MERCURY

In folklore and myth, Mercury's reign,

MERCURY

A trickster's domain, in mischief and gain.

MERCURY

A symbol of change, in sun and rain,

MERCURY

In alchemical fires, it rises again.

MERCURY

In the cosmos' vast expanse, it roams,

MERCURY

A messenger's journey, in celestial poems.

MERCURY

In twilight's embrace, where darkness looms,

MERCURY

Mercury whispers, in silent catacombs.

MERCURY

So raise a toast to Mercury's might,

MERCURY

In shadows and stars, in day and night.

MERCURY

A cosmic wanderer, in endless flight,

MERCURY

In the alchemist's vision, a guiding light.

MERCURY

For in the crucible of time, we find,

MERCURY

Mercury's essence, in heart and mind.

MERCURY

A symbol of transformation, intertwined,

MERCURY

In the alchemy of life, forever enshrined.

MERCURY

So let us cherish Mercury's tale,

MERCURY

In liquid silver, in myth and grail.

MERCURY

A celestial dance, where mysteries prevail,

MERCURY

In the alchemist's song, we set sail.

MERCURY

ABOUT THE CREATOR

Walter the Educator is one of the pseudonyms for Walter Anderson. Formally educated in Chemistry, Business, and Education, he is an educator, an author, a diverse entrepreneur, and he is the son of a disabled war veteran. "Walter the Educator" shares his time between educating and creating. He holds interests and owns several creative projects that entertain, enlighten, enhance, and educate, hoping to inspire and motivate you.

Follow, find new works, and stay up to date with Walter the Educator™ at WaltertheEducator.com

www.ingramcontent.com/pod-product-compliance
Lightning Source LLC
LaVergne TN
LVHW010412070526
838199LV00064B/5274